Leadership Development – Professional Skills to Catalyze Your Career

Topical Conference at the 2019 AIChE Spring Meeting
and 15th Global Congress on Process Safety

New Orleans, Louisiana, USA
31 March - 4 April 2019

ISBN: 978-1-5108-8683-4

Printed from e-media with permission by:

Curran Associates, Inc.
57 Morehouse Lane
Red Hook, NY 12571

CURRAN ASSOCIATES INC. proceedings .com

Some format issues inherent in the e-media version may also appear in this print version.

Printed by Curran Associates, Inc. (2019)

For permission requests, please contact AIChE
at the address below.

AIChE
120 Wall Street, FL 23
New York, NY 10005-4020

Phone: (800) 242-4363
Fax: (203) 775-5177

www.aiche.org

Additional copies of this publication are available from:

Curran Associates, Inc.
57 Morehouse Lane
Red Hook, NY 12571 USA
Phone: 845-758-0400
Fax: 845-758-2633
Email: curran@proceedings.com
Web: www.proceedings.com

TABLE OF CONTENTS

552663 Inclusion & Diversity: An Innovation Imperative

Fostering an inclusive work environment is a core business value at Chevron. Our strong belief is that an organization must welcome diversity of thought, life experiences and work styles to succeed – and an innovative esprit de corps is imperative to build on that success. Chevron Oronite President Bruce Chinn talks about the foundational role of inclusion & diversity in the innovation imperative.

551619 Building a Continuous Learning and Innovation Network

We live at time when globalization and technology are changing the world in profound ways: economic competitiveness, national security, and personal prosperity are being redefined. At the level of business, hiring companies cannot find people with the necessary skills for today's economy, from truck driving to data mining. Apart from the talent crisis, corporations must reduce new product cycle times and then deliver improved innovations to market, faster. Workers in every industry, meanwhile, find it more and more difficult to climb a career ladder, or even to get onto the bottom rung. For learners of all ages, the rapid evolution in technology and business practices accelerates and intensifies the need for new approaches to career awareness, preparation, and retraining. To help address both human capital development and the need for new knowledge universities and businesses must innovate and think "outside the box" to create new integrated models that lower cost; create customer value; provide a hands-on educational experience and build a foundation for continuous learning. Building a shared / local model that enables transition to the future state is an opportunity for professional societies. This presentation will cover work to date and how we can build globally shared practices – (1) career discovery, (2) integrated learning and (3) market to discovery to enable a new learning and innovation network

550594 Tools for Open Innovation

In today's global and fast-paced environment, the only way companies survive is to innovate. New products and services are required to keep pace with competitors, yet many executives feel that the ideas from within their organizations are stale and incremental.

Disruptive innovation, on the other hand, introduces new ways of solving old problems that engage and empower the customer. The most successful disruptive innovations are built on open innovation platforms, where we involve the customer in the design and development of the new product or service. Of course, there are challenges to managing these relationships, but when successfully implemented, co-creation leads to the highest levels of customer satisfaction.

In this presentation, we will discuss and practice some key tools for open and disruptive innovation. You will learn how to involve a customer in your product design without off-putting jargon or industry buzzwords. In particular, we will discuss the elements of *design thinking* to help you address customer needs and deliver innovations with more accurate product-to-market fit and speed-to-market.

Don't get left behind in a tidal wave of disruption. Learn how to apply open innovation tools to accelerate your success today.

550062 Innovating and Scaling up in Biomass Based Biorefineries – Lessons Learned

The first wave of scaling up biorefineries has encountered many failures and luck luster performance. Such an experience has cast a negative image on the whole industry. Lessons learned about the scaling up of a first of a kind Biorefinery processes will be presented and successful recipes will be discussed.

The open innovation practices of some industries will be compared with the proprietary innovation of the bio refinery industries and contrasts and benefits of the two will be compared.

550072 The Case for Open Innovation

Open Innovation is a term used extensively in companies of all sectors to encourage cooperation between companies, academia and government to reduce risk exposure and leverage capability in developing and commercializing technology. Shell has a decades-long history of open innovation in Shell GameChanger and continues today in support of its new formed New Energies business.

The process of managing open innovation does not follow the typical path in business which have historically developed technology with in-house capability and resources in order to extract as much value as possible. New ways of working and thinking in technology organizations must be adapted to leverage the possibilities of open innovation.

This talk will include generic case studies in which open innovation was successful and led to new products lines, and cases in which open innovation did not consider all business model or supply chain model aspects and failed to produce commercial success.

543895 Open Innovation Experiences: Delivering Success and Avoiding Pitfalls

Open Innovation Experiences: Delivering Success and Avoiding Pitfalls

Open Innovation with a strategic partner can bring your technology to the market quickly and efficiently. However, if not managed well it will lead to project stagnation and missed opportunities. MATRIC has collaborated in over 200 projects with over 100 different customers. The projects have had varying degrees of technical and commercial success, and the learnings highlight key considerations in executing open innovation development opportunities.

I will discuss three projects that used three different approaches to open innovation and how these impacted the project execution and success. The three projects had several similarities:

- the early stage technology discovery work was completed before contacting MATRIC
- the customers wanted to maintain full ownership of the intellectual property
- they didn't have the resources (people and/or facilities) that could be used to develop the technology
- they had external stakeholders in the success of the project who were not engaged directly in the technology development but did contribute time, money or resources to the development effort.

The first customer is a mid-sized company developing a specialty product that would be differentiated in the market by its performance, and could have additional value to the company through the use of lower cost raw materials than their current alternatives. Their open innovation challenge was to reduce the process costs, both operating and capital costs, primarily through an increase in the catalyst productivity. The customer had identified the molecule of interest internally, but did not have the catalysis skills and process development capabilities to prepare for commercialization. Additionally, the target consumers of the product would need to be involved in the final qualification. MATRIC formed a joint technology team with the customer that included the catalyst and process expertise of MATRIC with the product development team from the customer. The team worked very closely together, with weekly meetings and frequent visits by the customer to MATRIC labs to observe the progress and discuss the results. The team was very successful, achieving a 100-fold increase in catalyst productivity well ahead of the project schedule. The consumer of the final product was engaged partially throughout the work, but a critical product test was not completed until late in the technology development. The test results indicated that product needed to be reformulated to avoid adverse effects. This resulted in reworking the catalyst development , with the resulting increased development costs and delayed implementation.

The second customer was interested in developing technology that would create new markets for its product. Earlier work at the laboratory bench scale had demonstrated a technology that could be used to produce an existing commodity product, however the process economics needed to improve. A team was formed of MATRIC experts in process chemistry and process development, along with representatives of the customer and experts familiar with the customer's suppliers and their related technologies. The team had virtual meetings biweekly with periodic face-to-face meetings that included the team members and representatives of the customer's suppliers. The project started with a techno-economic assessment to understand the critical areas for technology development that will lead to project success. It has developed from a bench scale to a larger-scale laboratory study to better understand some of the process challenges and prepare for pilot plant demonstration of the technology. The team communications have allowed the project to progress rapidly, but in a disciplined manner, and to be ready for a smooth transition to the next phase of the technology development.

The final customer was interested in leveraging their technology to create a new path to an existing product for a key consumer. Although their process was a key linchpin in the manufacturing process for the ultimate product, process discovery and development was needed to produce the final product. The customer worked with another research organization to develop the process on a bench-scale, but needed more capabilities to scale the process to a pilot plant and produce large market development samples. The customer shared with MATRIC the reports from the other process development organization and looked to MATRIC to design the pilot plant facility from these reports. Clarifications of the technology were handled by the customer, who would occasionally contact the previous research organization when necessary. This slowed the transfer of the technology and eventually led to a late learning that significantly impacted the process design, requiring redesign of whole sections of the pilot plant with significant increases in the engineering and construction costs. The delays and increases impacted the market development opportunity, and eventually the project was placed on hiatus.

The key learnings from these projects are:

- It is important that in open innovation the customers and research organization must form a partnership with close communication
- All key stakeholders need to be represented in the project to make certain that the process development delivered meets all expectations
- Technology transfer between project stages needs to be thorough. This is readily achieved by having continuity within the process development team, and especially if the open innovation partner is capable of carrying the project through the multiple phases.

557365 Open Innovation Model Successes-New Process in Spray Drying

A success story regarding Open Innovation techniques employed by Pulse Combustion Systems will be presented along with case studies.
We start with a brief background of our technology, followed by a description of our recent focus on solving customer problems via Cooperative R&D. We will present lessons learned and will also show how the company is growing rapidly as a result of using Innovation in a proactive and collaborative manner to solve customer problems. Four cases will be presented to highlight the successes and challenges of the Open Innovation Model.

551560 Ethics for Registered Engineers, a 1PDH Professional Ethics Presentation

LAPELS staff will discuss the Board, enforcement activities and investigative procedures, recent cases, the duties of a firm's supervising professional, ethical issues and other common questions.

551629 Ethics As an Underpinning to Safety

Ms. Grubbe will demonstrate and define ethics in support of good safety, and will show how ethics is eroding in academia, government and in industry. She will also suggest a way for academics and university industrial board members to consider inserting ethics into every engineering curricula.

551630 Case Study Reviews in Ethical Failures Contributing to Process Safety Incidents

Ms. Grubbe will use her knowledge and experience of several serious process incidents, and will tie the messages of the ethical basis of safe operations to key instances where ethical failures occurred. She will discuss the hard dynamic that occurs when an engineer chooses to remain silent vs. speak out about a noted issue.

551893 Ethical Issues in Safety- Optimizing Safety Culture and Leadership to Rise Above the Challenge!

In today's fast paced and demanding business environment engineers and operators are faced with competing priorities and pressures. During these complex times, everyday decisions must be made without compromising the Ethics of Safety and allowing safety priorities to rise to the top . During this session, the importance of a strong safety culture and leadership behaviors will be discussed as key to successfully navigating the Ethics of Safety. Practical short and long term solutions will be shared that can be applied across a broad spectrum of industries.

550530 Building Your Career Army

When it comes to career advancement and planning you often hear about networking. Most people think of networking as making contents, attending networking events or grabbing one lunch with someone. Most people also think networking only matters when you are looking for a job. Networking, however is more strategic and more involved than just that. It's about building an army of people in your career ecosystem that can help you strategize, execute, and fight for your career day to day. In this session, we will talk about what a career army looks like, who you need to enlist, and how to mobilize your troops to help you advance your career.

550559 Mentoring Vs Coaching, What Is the Difference? What Is the Benefit of Each?

Everyone thinks they need or want a mentor. You actually may need a coach or both a coach and mentor. You may think there is little difference between coaching and mentoring. You may think they are similar or that they are the same. They are different and we will go over those differences. When do you need one vs the other. We will also go over the tips for successful interactions in a mentor, mentee relationship. Roles and responsibilities of the mentor & mentee will be looked at. Plus a personal view from the trenches of having both successful and non successful coaches, mentors, and mentees.

552664 Innovation and Emerging Technology Perspectives in Hydrogen and Fuel Cells at the U.S. Department of Energy

Today the technology around generating efficient and sustainable energy is rapidly evolving and hydrogen and fuel cells are emerging, versatile examples within a portfolio of options. Dr. Sunita Satyapal will provide an overview of the U.S. Department of Energy's (DOE) activities within the Office of Energy Efficiency and Renewable Energy (EERE) focusing on hydrogen and fuel cell technologies. This is an exciting time in the fuel cell industry with several thousand commercial fuel cell vehicles now sold or leased, hundreds of fueling stations worldwide, and more than a quarter of a million stationary fuel cells providing clean, reliable power. DOE's Fuel Cell Technologies Office (FCTO) addresses key technical barriers faced by hydrogen and fuel cells through a comprehensive portfolio of early-stage research and development (R&D) with the potential to meet technical and cost targets that enable competitiveness with incumbent technologies in the market.

The presentation will cover DOE's H2@Scale initiative which will enable innovations to generate cost-competitive hydrogen as an energy carrier, coupling renewables, as well as nuclear, fossil fuels, and the grid, to enhance the economics of both baseload power plants and intermittent solar and wind, enhancing resiliency and avoiding curtailment. Hydrogen can be stored, distributed and used as a fuel or feedstock in transportation, stationary or multiple industrial sectors, while creating an additional revenue streams. The presentation will also provide examples of safety, codes and standards R&D activities that FCTO is funding to enable the H2@Scale initiative. The talk will also provide a preview of the new Center for Hydrogen Safety through a collaboration between American Institute of Chemical Engineers and DOE's Pacific Northwest National laboratory, as well as multiple stakeholders worldwide.

553087 Experience with Starting and Operating a Consulting Business

In this presentation I will share my experience in starting and successfully operating a consulting business in environmental, health and safety. I started a consulting business ten years ago and since
have managed to successfully generate increasing revenue stream while maintaining good profit margins.

This talk will discuss how to carve a niche which has a demand in service business, trials and tribulations
in starting an enterprise from scratch including generating initial revenue, obtaining appropriate insurance and resources required both financial and human. I will also discuss my experience in setting
up company name and logo.

Maintaining and sustaining a small service business requires being both a" seller" and "doer". I will discuss balancing between need to sell work as well as complete technical projects. I will discuss my own
overall approach to generating steady revenue stream. I will relate my experience in maintaining cash
flow, balance sheet and other financial issues.

I will discuss some of the attitude and aptitude required to be successful in business based on my experience. There is a lot of satisfaction and a sense of accomplishment in operating a business if there
is a methodical approach with right attitude and ability to adapt to changes.

Turning a Consulting Idea into Reality

Jack Hipple, Principal

TRIZ and Engineering Training Services LLC
Tampa, FL

Abstract

Starting a consulting business is not as easy as printing business cards, posting a web site (though both are important), and waiting for the phone to ring. It involves networking, contacting friends and associates, giving pro-bono presentations at both local and national meetings, publishing, and using resources such as LinkedIn.

This presentation will review the steps needed to establish a consulting and training services consulting business from scratch and emphasizing the importance of professional association participation in turning an idea into reality.

Consulting: Is It For You?

After a 30 year industrial career, focused on both basic chemical engineering (primarily at Dow) and materials (National Center for Manufacturing Sciences, Ansell Edmont, Cabot), I was exposed to a rather unique problem solving technology known as TRIZ ("Theory of Inventive Problem Solving") and decided to look for an opportunity to work in this area. After a brief time with one of the original Russian emigrant-based consulting companies to train and learn, I decided to start my own business. Having never done this before, I encountered a number of challenges, as well as opportunities, that will be shared. They may be of some assistance and interest to those considering starting their own business.

Practical Issues

Everyone has boundaries that affect their decisions about starting an independent business and there were mine:

1. Desire to grow a business into a substantial enterprise. Many people starting consulting businesses have a goal of growing their business into multimillion-dollar enterprises, requiring hiring other employees, needing an accounting and marketing structure, and requiring decisions on employee benefits. My intention was simply to keep myself fully employed to the extent necessary to maintain a modest lifestyle.
2. Linkage with major consulting organizations. This can be a stated objective, but again, was not on my list of priorities.
3. Health insurance. The level of this concern varies greatly from state to state and is obviously affected by family size, spousal employment, and personal health issues. Given my personal health issues, this was a significant concern.

4. Family. As empty nesters with all children having completed college and employed, and our 4 daughters all married, medium term financial issues were not of concern. Obviously, it is easy to imagine other situations with college bills, teenagers at home, etc. that would have great significance.
5. Spousal support and feelings. As mentioned before, short term financial issues can be of great concern. Not all married couples see financial risk the same way and it is critical to have a frank conversation about the fact that there may be no income for several months. If a spouse is working and children are still of school age, then issues relating to child care will come up, especially if the nature of the consulting business requires travel on short notice.
6. Finances. Two key points here. First, the health issue mentioned previously. If a previous employer will or cannot allow insurance continuation (usually as a much higher cost), and a spouse is not employed by an insurance providing organization, this can be a major cost issue that needs to be considered and is part of the financial planning
7. Unpredictable events. These could be health issues but can also be unpredictable external events. Case in point was the 9/11 terrorist attack. From personal experience, and that I know of others, unless you were a security consultant, the phones stopped ringing. No one knew what was going to happen and businesses froze in place without making commitments of any sort. This can be a short-term financial crisis, and in my case, required my wife to go back to work in order to be able to maintain health insurance.
8. Travel. This again is a highly personal, and family connected decision. There are many successful consultants who do no or minimal travel. This is more common in a major metropolitan area, but that area must contain industry that has some overlap with your intended consulting focus. If heavy travel is involved, the spouse is working, and children are of school age, this can be a major, and sometimes costly, issue to deal with.

Suggestions for How to Make it Work

Since, in my case, the goal was to grow business to fully employ myself and not grow a large company, and my resources were rather limited, here were the low-cost items that allowed my business to grow into a success:

1. Business cards. Unless I know someone has died, I NEVER throw away a business card I have collected at a meeting. It costs virtually nothing (except an Email account) to send appropriate Emails to past business acquaintances telling them you have started a new business around something unique or better yet, make a phone call (a long-lost art). And after you have gotten a job or two and know you will be in a certain area over a known period, ask the right people about getting together for lunch or dinner, possibly extending the main purpose business stay for a day or two.
2. Local societies. I will address national societies next, but in almost any reasonably sized city, there are local chapters of every national engineering, business, and technical

society. And there are numerous chamber of commerce and other such groups who are surprisingly eager for speakers. It's not that the groups don't have potential speakers, but it is far more likely that a consultant has and will take the time to prepare a presentation. Paying a meeting fee of less than $50 almost always leads to follow up contacts and potential business.

3. National societies. I teach two courses for both AIChE and ASME, one in Chemical Engineering for Non-Engineers and the other in the TRIZ problem solving process. These came about, in part, by coincidence, but point out the value of maintaining a relationship, and actively participating in such societies and in their sub-groups, such as AIChE's Management Division. 15 years ago, I tried to make contact with AIChE's training group but was not terribly active in AIChE, so did not have a personal contact. I became resigned that the connection would not yield anything, but just as I was giving up the idea, I saw a course listing for TRIZ Inventive Problem Solving (my specialty) listed in the AIChE training catalog. My energy level increased to the point where I persisted, and finally talked to the training manager at the time, who was very cordial and receptive and told me that AIChE always wanted to have a back-up instructor for any course offered and gave me the opportunity to teach the next TRIZ class without ever having met me! As any good training manager would do, she came in and out of the room over the 3-day period, and at the end of the course, offered me the opportunity to be the primary instructor with no intentional switching.

   And then the next great thing happened when I was told something like, "I bet you could teach anything". After joking a bit and telling her that I was not a TV pitchman, I asked what else she had in mind. AIChE needed a backup instructor for its "Chemical Engineering for Non-Chemical Engineers" course, due its current instructor having a growing unpredictable eldercare issue that might prevent her fulfilling a commitment to teach. Though 10+ years away from having done serious ChE calculations, the idea intrigued me under the condition that I be reimbursed for going to the next public class and sitting in the back of the room and watching, after which I would decide. In the mean time I was sent a copy of the course materials to review. They did not scare me as much as I thought, so I packed my bags and got ready to fly to New Orleans on Monday morning. Late Friday afternoon, the phone rang, and I was told that the current instructor's mother had passed away and that I would be teaching the class. You can only imagine what went through my mind! I have been teaching the course ever since.

   My message here is that my major source of income, requiring little or no marketing expense on my part, has come from interactions with AIChE, thought the TRIZ class emigrated over to ASME during AIChE's financial crisis a while back. And these public classes inevitably lead to additional training opportunities, requiring little or no marketing expense on my part.

Summary

My summary of advice, given goals and situations similar to mine, is thus:

1. Make sure consulting is what you want to do
2. Focus what kind of consulting you want to do and base it on your strengths and uniqueness.
3. Make sure you and your spouse agree on this and how it will affect your lives, both personally and financially. Understand how it may affect family life. Don't leave children out of the discussion.
4. Use your past connections. Don't throw past information and resources away.
5. Prepare for 6 months of no income and consider what you will do if some unforeseen issue arises that stops your income for several months.
6. Use your local, and more importantly your national societies, to advance your business ideas and contacts.
7. NEVER throw away a business card unless you can match it with an obituary.

553059 Work in Retirement Can Yield Numerous Benefits

Several factors can induce a worker to opt to retire from the daily routine and grind of specific work assignments and requirements. Some could be health factors, family obligations, travels and fun, and the desire to impact several companies and professional workers beyond the confines of the current work sphere.

This presentation may be of interest to those who are in "retirement", to those who are contemplating "retirement", and to those who wish to continue producing and impacting companies, and others during retirement. I will be sharing with you the joys and benefits that can be derived from running a technical, engineering and optimization consultancy services.

In founding and working for Prafis Energy Solutions, I have had the time to meet family obligations. It has given me the time to re-energize my growth as I continue on my spiritual journey. "Retirement" has enabled me to work on a number of interesting projects, manage health care challenges, and welcome the recent increase in the number of my grandsons from 3 to 5. The usual travels have been done, and more recently, my book with the title of Catalytic Naphtha Reforming Process was published in 2018 by CRC Press of the Taylor Francis group.

552685 Fishing for a Solution: A Hands- on Approach to Continuous Improvement

Key to any continuous improvement effort is having the ability to problem solve your way through the obstacles you encounter. This workshop will give hands-on practice in two techniques that teams can use in problem solving. Join us in this 60 min interactive workshop in exercises you can take back to your teams.

552405 On a Unit Level: Managing Operational Excellence

Continuous improvement is something that is managed at many levels, at a company level, site level, and unit level. This presentation will introduce how it is managed on a unit level or at a lower level where people not only have to buy off on using continuous improvement, but have to use it on a daily basis.

552694 Workshop: Managing Yourself

This session will be an **interactive workshop** that will be split into 2 parts:

1. Auditing yourself and your career: how to do a self-assessment on what skills you have, what type of role you want to have, and what skills you need to develop to get there

2. Branding yourself professionally: how to create a professional brand and market yourself

Oftentimes, a new role signifies a promise of a new future but how often do you take the time to plan out that role to prepare you for your next one? This workshop provide ideas to help you position yourself for your next role. It will also teach you how to recognize and articulate your transferable skills.

551928 Soft Skills for Young Professionals

Going from being a student to working is a big transition for most of us. It requires us to learn how to navigate the real world, and we can use all the help we can get to do it well. Soft skills are the most important secondary skill (the primary one being the vocation that you are being employed for) that can help us here, and not only in our careers but also in personal lives. This session will focus on some of the necessary soft skills regardless of the degree we have, reflect on which soft skills are transferrable and which need to be developed in order to have a rewarding time at our organizations.

551786 The *Trusted Advisor*: A Model for Career Success

The classic book *Trusted Advisor* by David H. Maister presents highly effective tactics for building and maintaining long lasting relationships as part of a successful career strategy. A Trusted Advisor is a person who has purposefully elevated their personal and professional relationships to a level where solving problems is more valuable than technical mastery. Trusted Advisors have self-confidence and driven curiosity. They believe that there is distinction between business life and personal life, but both are seamlessly interconnected.

This presentation will outline the three core skills needed to become a Trusted Advisor: earning trust, giving advice and building relationships. Effective ways to develop each of these three skills will be presented in an easy-to-understand format. Pitfalls for destroying trust will be discussed along with tactics to avoid doing so. The goal of the presentation is to outline these tactics in the form of constructive actions that you can perform daily for the rest of your career. Trusted Advisors are easy to spot: they seem confident, selfless and laser focused on helping themselves and others achieve their goals.

550124 Networking Strategies to Propel Your Career

This presentation discusses various early career strategies to build business relationships with coworkers, clients, and vendors. Develop a strong network now for any future downturn or advancing within your company.

550521 Engineering Creative Thinking

What do you think of when you hear the word Creativity? If you're like me, you often think of something artsy that you would see on Instagram or Pinterest. You also probably think creativity is a trait people are born with or without. You probably also think creativity is something you don't have to utilize in your technical jobs. However, creativity and creative thinking is a skill that can set you apart in any job. Creativity isn't about originality or thinking outside of the box as much as it is about learning how to connect ideas, growing a bigger box, and building your mental inventory of knowledge. In this session we will go over practices, tools, and tricks for how us engineers can expand our creative thinking to make us more innovative and strategic in how we approach our problem solving and our jobs.

552665 Integrated Safety Management - Taking Safety Performance to the Next Level

Gavin Towler, Vice President and Chief Technology Officer, Performance Materials and Technology, Honeywell, will provide an overview of recent advancements in risk assessment and a look ahead at the future direction for chemical and process industries. Today, risk assessment techniques play a fundamental role in process design, implementation of safety systems, maintenance, and operation management, as well as an essential tool for the development, operation and expansion of process installations. Mr. Towler will discuss the latest dynamic risk assessment approaches that support real-time decision making.

AUTHOR INDEX